Serdyuk Oleksiy, Shkurat Natalia

Space exploration

THE EARTH'S ORBIT

MWT
2018

SPACE EXPLORATION. THE EARTH'S ORBIT
by Serdyuk Oleksiy and Shkurat Natalia

This book will tell you about how humanity has made its first steps in conquering outer space. About how the first rocket took off, how the first spacecraft sent its signal to Earth, how people flew into space, how different countries, such as the USA, the European Union, China, Japan, India, Russia, master space technologies, and as well as many other things related to human activities in the Earth's orbit.

Copyright © 2018 by Serdyuk Oleksiy, Shkurat Natalia

ISBN-13: 978-1719085076

Photos by NASA, SpaceX, ESA

Cover photo:
International Space Station and the docked space shuttle Endeavour (by NASA)

Send Questions, Comments and Feedback: mwt.publishing@gmail.com

CONTENTS

- THE BEGINNING 4
- SATELLITE 8
- THE FIRST STEPS IN COSMONAUTICS 12
- LAUNCH ROCKETS 24
  - SEA LAUNCH 28
  - FALCON 29
  - MANNED FLIGHTS 32
  - SPACE TOURISM 36
  - CARGO SPACECRAFT 38
- ORBITAL STATIONS 40
  - INTERNATIONAL SPACE STATION 44
  - SCIENTIFIC RESEARCH IN SPACE 47
- SATELLITE SYSTEMS 48
  - METEOSATELLITES 49
  - SATELLITE CONNECTION 50
  - SCIENTIFIC SATELLITES 52
  - HUBBLE SPACE TELESCOPE 54
  - INTELLIGENCE SATELLITES 58
- SPACE AGENCIES 59

THE BEGINNING

Scientists' minds have been excited by dreams of flying in the sky and conquering outer space for many centuries. Interplanetary travels were often described in the books of science fiction writers.

In 1881, M.I. Kibalchich formulated the idea of a rocket aircraft. In 1883 the idea of using rockets to reach outer space was developed by a scientist, Konstantin Tsiolkovsky. He even drew a diagram of a spacecraft. And in 1886 he formulated the theory of jet propulsion. In 1903, he published an article theorizing that a rocket would be able to go into space. Later, he developed his ideas and calculated the second cosmic velocity (escape velocity). Tsiolkovsky became one of the founders of the new science – Cosmonautics.

M. I. Kibalchich Ukrainian inventor, a rocket pioneer

Tsiolkovsky's ideas were picked up and developed by a German researcher Hermann Oberth.

Oberth came to the conclusion that it is possible to travel into space by using a multistage rocket with liquid fuel. He justified the possibili-

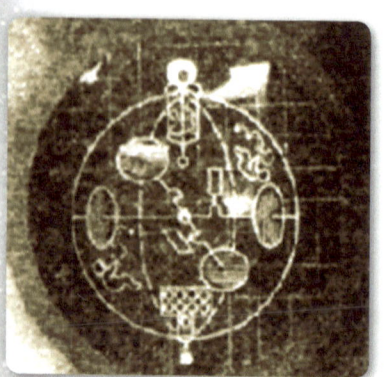

Draft of first space ship by Konstantin Tsiolkovsky, Manuscript "Free space", 1883

TSIOLKOVSKY KONSTANTIN EDUARDOVICH
1857-1935

Russian scientist, philosopher and writer. Author of works on aerodynamics and aeronautics. One of the founders of Cosmonautics. Tsiolkovsky had been dealing with issues of the movement of jet vehicles since 1896. He developed schemes for the operation of rockets for space travel. He was a supporter of the idea of colonization and exploration of space.

ty of creating such a rocket with great accuracy. Oberth shared his ideas with an American scientist Robert Goddard, who worked on developing a liquid rocket engine. In March 1926 Goddard managed to complete a working prototype of the engine and he launched the first rocket powered by liquid fuel.

The theme of space flight was becoming more and more popular throughout the world. The results of the works by Tsiolkovsky, Oberth and Goddard were used by scientists from many countries.

V-2 rocket on Meillerwagen near Cuxhaven in 1945

Space exploration

The deepest interest in rocketry was in Germany. In 1936 a rocket center was founded in Peenemünde.

An engineer, Wernher von Braun, was appointed as one of its leaders. Under his leadership, a ballistic rocket A-4 was developed in that rocket center. In serial production, the rocket was named V-2. That rocket was the first man-made object

A V-2 launched from a fixed site in summer 1943

A launchpad at Peenemünde as depicted in a miniature at the Deutsches Museum

The Earth's orbit

HERMANN OBERTH
1894-1989

Austrian-German physicist, researcher, inventor. Founder of rocket technology and astronautics.
He was engaged in researches in the field of developing rockets and launching them into outer space. He was the first to formulate the idea of creating an orbital telescope. In the 1950s, he was dealing with the problems of space exploration by people.

in history to reach outer space. In 1944, several vertical rocket launches were made. It reached an altitude of 188 km.

After the end of the Second World War, the USSR and the United States gained access to the rocket technology of the Third Reich.

Rocket site at Peenemünde, 1943

SATELLITE

The Second World War gave a powerful impetus to the development of military and civilian technologies. In the early post-war years, Cosmonautics occupied one of the key places in scientists' research.

After the Second World War, scientists and inventors did not abandon their plans to explore outer space. Both in the USA and in the USSR, engineering groups working in the field of rocketry had been promoting and developing ideas since 1946 for launching the first space satellite of the Earth. In the Soviet Union, the ideologist of this project was the Ukrainian design engineer Sergiy Pavlovich Korolev, and in the United States – the German engineer Werner von Braun.

**KOROLEV
SERGIY PAVLOVICH**
1906-1966

Ukrainian scientist. Constructor. The largest specialist of the twentieth century in rocket engineering and astronautics. The creator of the Soviet space industry. Ideas and development of Korolev's design office made it possible to launch the first artificial Earth satellite, launch of the first cosmonaut, first spacewalk and much more.

Mikhail Tikhonravov and his team

On May 12, 1946, von Braun's engineering team submitted a report to the US Department of Defense, which described the possibility of putting an artificial earth satellite into orbit by 1951. On May 21 of the same year, Mikhail Tikhonravov made a similar report to the USSR Ministry of Aviation Industry. However, neither of the engineering groups received support at the state level.

In October 1954, the Organizing Committee of the International Geophysical Year sent a request to the leading countries to consider a project to launch an artificial Earth satellite for research. US President Dwight Eisenhower responded by saying that the United States was ready to implement that project. As the relations between the US and the USSR were imbued with a spirit of rivalry, a similar announcement was soon made by the Soviet leadership.

PS-1 in laboratory

In the USSR, the date for the satellite launch was set for 1957. The satellite which was prepared for the launch in the technical documentation was listed

Space exploration

The design of the first artificial Earth satellite. There is a transmitter and a power source inside.

as Sputnik-1 — the simplest satellite. On October 4, 1957, the R-7 intercontinental ballistic rocket launched the first artificial satellite into the Earth's orbit.

The satellite was in orbit for 92 days — until January 4, 1958. It made 1,440 revolutions around the Earth. On Earth, information about the launch of the satellite caused a sensation. People in different countries all over the world started to watch the sky at night, trying to see among the many glowing points the one that moved around the Earth.

This is how people began to master outer space.

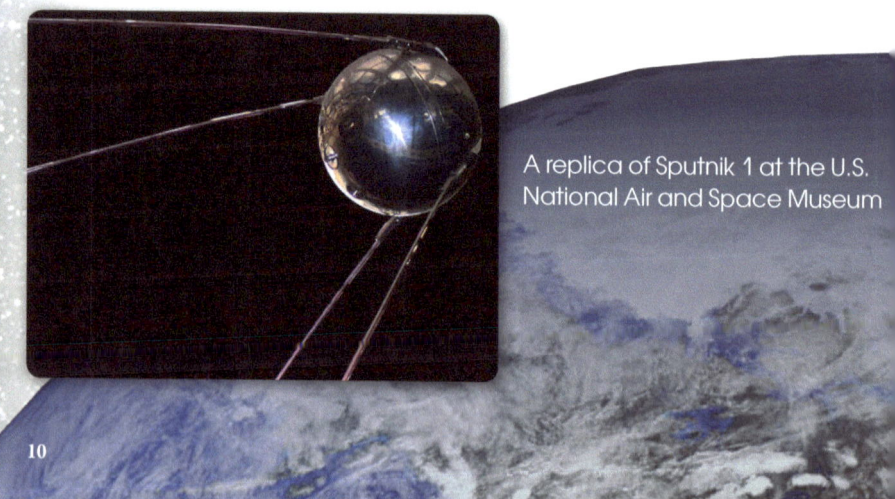

A replica of Sputnik 1 at the U.S. National Air and Space Museum

Real-size replica of Sputnik-1 in Madrid (Spain)

© Luis García

THE FIRST STEPS IN COSMONAUTICS

The launch of an artificial satellite made space exploration one of the most prioritized branches of science. In the early decades, space programs were developing very actively.

After the launch of the Soviet satellite, the United States accelerated the process of creating its own satellite. So the space race began and it lasted 18 years.

On November 3, 1957, the Soviet Union launched the second satellite, which weighed 508 kg. It was effectively an entire self-contained space laboratory. In addition, it had a dog on board – Laika, the first living creature sent into orbit. However, after several orbits around the Earth, the dog died because of overheating. But the launching of a living creature into the cosmos brought scientists

The dog Layka on board of Sputnik-2

Monument to the dog Layka

WERNER VON BRAUN
1912-1977
German design engineer, who worked in the United States. Creator of ballistic rockets. Founder of the US space industry.

substantially closer to the possibility of sending a human being into space.

In the United States, the Navy Research Laboratory (the Vanguard project) and the RAND Corporation Laboratory (the Explorer project), headed by Werner von Braun, worked on the artificial satellite project. Priority was given to the project Vanguard by order of the state.

On November 11, the United States announced the launch of its satellite, Vanguard. But that launch ended in complete failure. The United States had to return to Von Braun's project

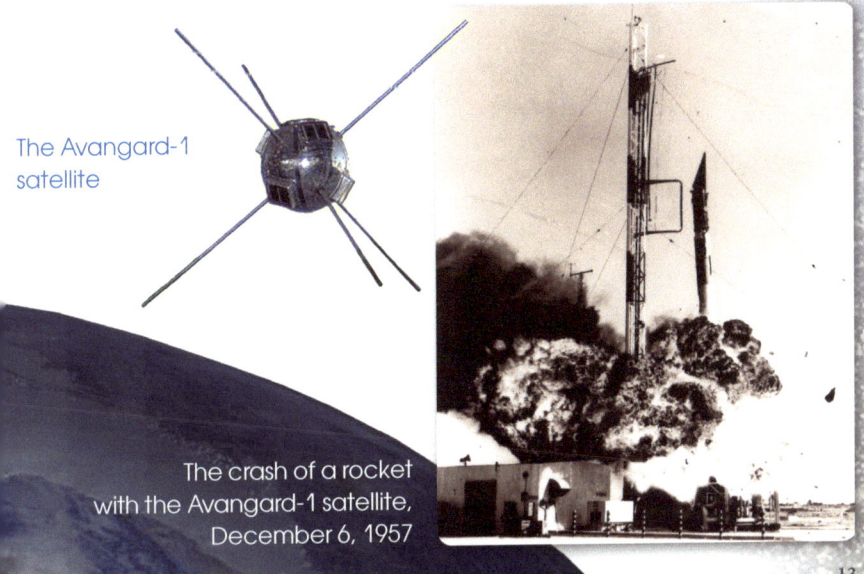

The Avangard-1 satellite

The crash of a rocket with the Avangard-1 satellite, December 6, 1957

Space exploration

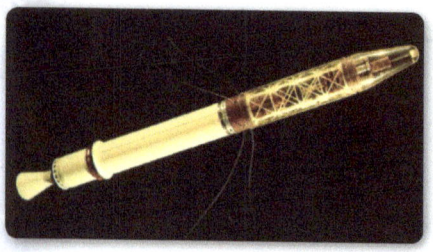

Explorer 1

Explorer. The launch of Explorer was successful. It was held on January 31 1958. The United States became the second space power.

Further exploration of near-Earth space was very intensive. On May 28, 1959, the United States launched a monkey into space (AM-18 mission). And the Soviet Union sent the station Luna-1 towards the Moon, which passed at a distance of 6,000 kilometers from it and entered a heliocentric orbit. In the same year, the USSR sent two more stations towards the Moon. The first of them, Luna-2, reached the surface of the Moon, and the second, Luna-3 was the first to study the dark side of the Moon, invisible from Earth, and it also practiced a gravitational maneuver.

On August 19, 1960, the Soviet Union launched dogs Belka and Strelka into space on the ship Sputnik-5. On August 20 they successfully returned to Earth.

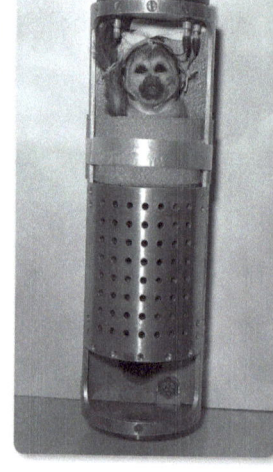

A monkey, Miss Baker, in bio-pack couch being readied for Jupiter (AM-18 flight).

The 60s were marked by the appearance of new space powers on the map. In 1962 the United Kingdom and Canada launched their first satellites, in 1964 – Italy, in 1965 – France.

Vanguard II is an Earth-orbiting satellite launched February 17, 1959

Launch of Explorer 1

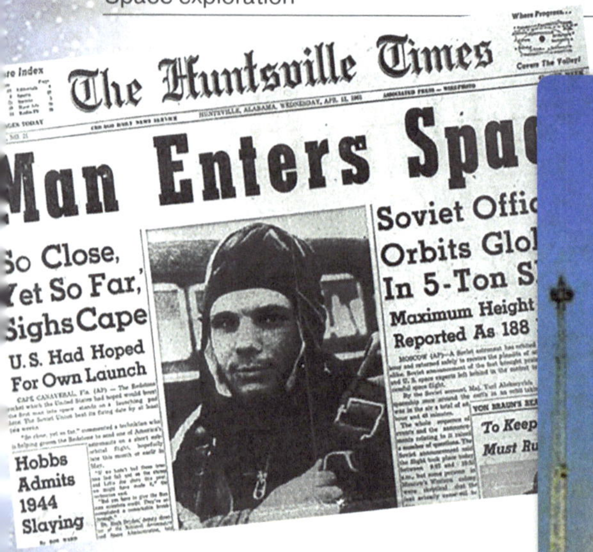

The Huntsville Times

Man Enters Spac

'So Close, Yet So Far,' Sighs Cape

U.S. Had Hoped For Own Launch

Soviet Offic

Orbits Glol

In 5-Ton S

Maximum Height

Reported As 188

To Keep

Must Ru

Hobbs Admits 1944 Slaying

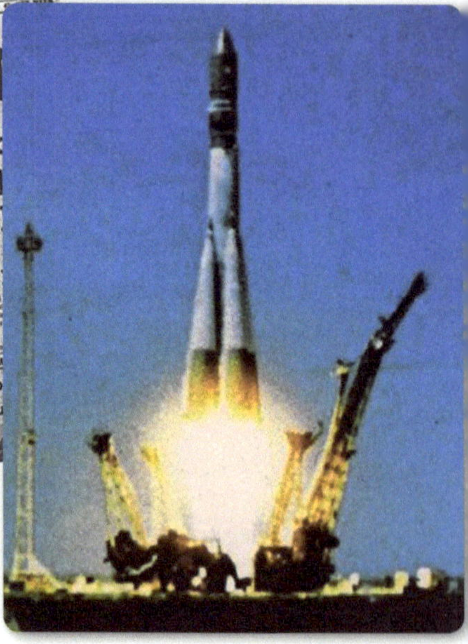

The launch of the Vostok spacecraft with Yuri Gagarin on board, April 12, 1961

On April 2, 1961, one of the most significant events in the development of Cosmonautics took place. For the first time, a spacecraft went into space with a human being on board. Pilot-cosmonaut Yuri Gagarin made the first flight to outer space on the Vostok spacecraft.

The descent module of the Vostok spacecraft after landing

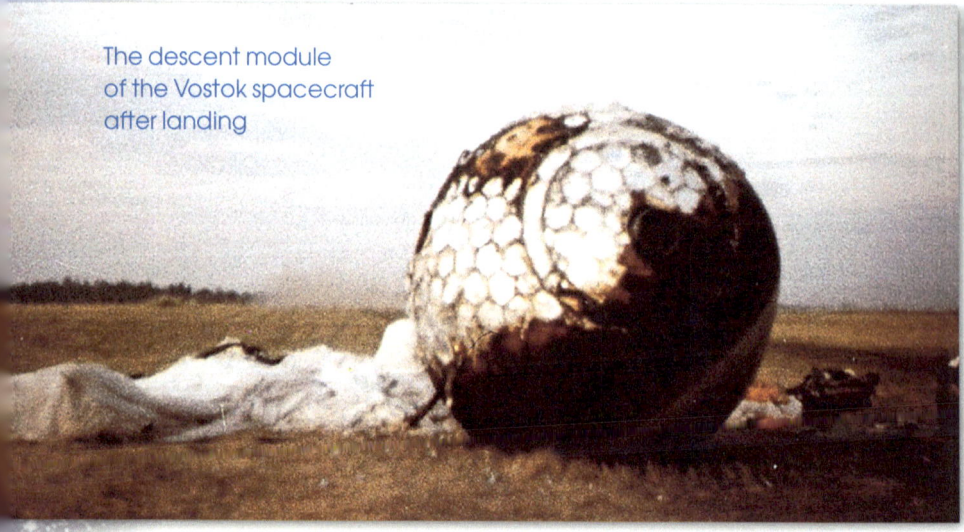

GAGARIN YURI ALEKSEEVICH
(1934-1968)

Pilot-cosmonaut. On March 3, 1960, he was selected for a group of astronaut candidates. On April 12 1961, he flew into space. He was an honorary member of the International Academy of Astronautics. In 1968 the International Aviation Federation established the Gagarin Medal, which is awarded to people who have made a special contribution to the development of space and aviation.

The Vostok flight was conducted in automatic mode. The cosmonaut served as a passenger. But the program also provided some manual control if necessary. In reality, at that time there was no information about how a person would behave in a state of weightlessness, how stable his mental state would be. Therefore, the transition to manual control required entry of a special code, which could only be entered by a sane and rational person. The envelope with the code was on the control panel.

Yuri Gagarin in the cabin of Vostok

"When I orbited the Earth in a spaceship, I saw how beautiful our planet is. People, let's take care of this beauty and multiply it, don't destroy it!"
Yu. A. Gagarin

After Gagarin's flight, the possibility of human beings staying in outer space was proven in practice and, as a consequence, the way was cleared for further experiments in this direction.

On May 5, 1961, the American Redstone-3 rocket launched the spacecraft Freedom 7, carrying the American astronaut, Alan Shepard.

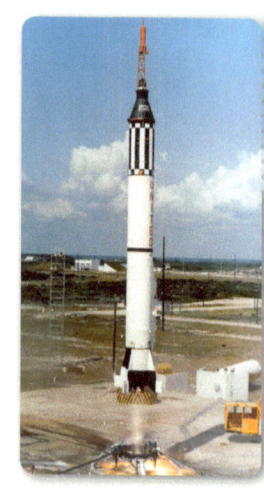

Launch of the Mercury spacecraft with Alan Shepard on board

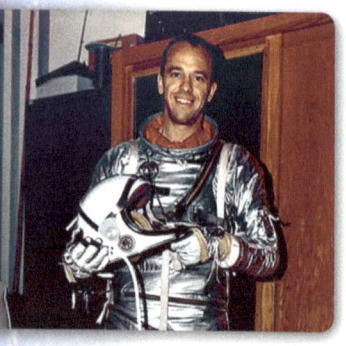

ALAN SHEPARD

1923-1998

American astronaut. Military pilot. The first American to fly into space. The second flight was performed during the mission that landed on the moon — Apollo 14. At that time, Shepard was 47 years old. He was awarded gold medals by NASA.

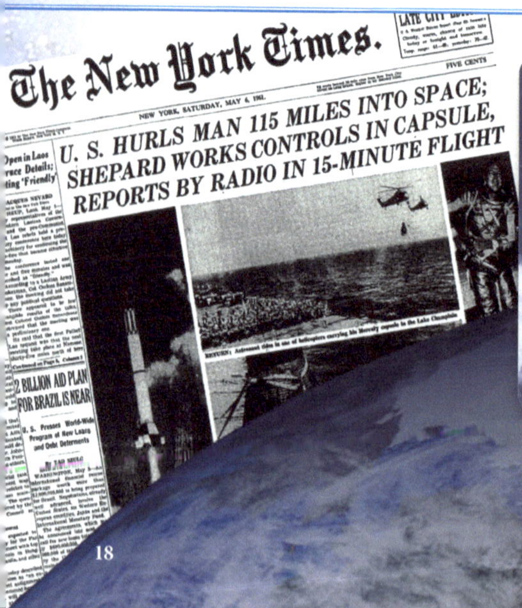

The New York Times.

NEW YORK, SATURDAY, MAY 6, 1961.

FIVE CENTS

U. S. HURLS MAN 115 MILES INTO SPACE; SHEPARD WORKS CONTROLS IN CAPSULE, REPORTS BY RADIO IN 15-MINUTE FLIGHT

$2 BILLION AID PLAN FOR BRAZIL IS NEAR

U. S. Presses World-Wide Program of New Loans and Debt Deferments

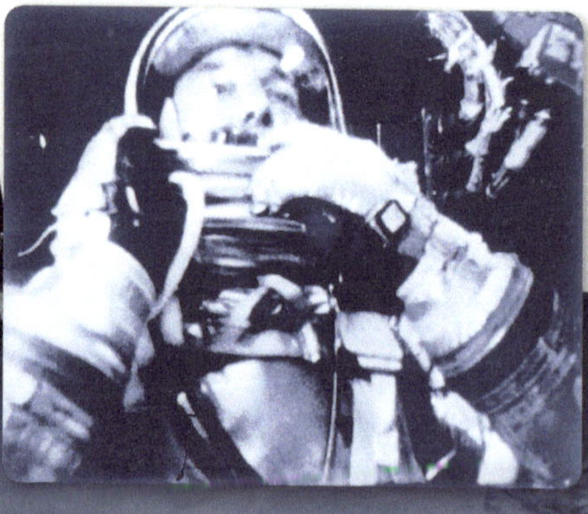

Alan Shepard on board Mercury

The Earth's orbit

On August 6, 1961, in the spacecraft Vostok-2, the Soviet cosmonaut German Titov spent more than 24 hours in orbit — 25 hours and 18 minutes, which was the first time in history.

And on June 16, 1963, on the Vostok-6 spacecraft, the first woman cosmonaut Valentina Tereshkova went into outer space.

Valentina Tereshkova

On October 12, 1964, the first multi-seat ship Voskhod-1 was launched into space, the crew of which consisted of three people.

On March 18, 1965, a cosmonaut A. Leonov (Voskhod-2) was the first ever to go into outer space. A month and a half later, on June 3, an American astronaut Edward White (Gemini-4) went into outer space.

Edward White

Alexey Leonov

Launch of the carrier rocket with the Voskhod-2 spacecraft

E. White in outer space

Space exploration

Launch of Gemini 4

The rivalry between the two countries in the space industry led to a breakthrough in the development of technologies in outer space exploration. But it also had some essential shortcomings. Because of the reduced deadlines for the implementation of projects, there was a lack of time to ensure the safety of flights. Therefore, unexpected glitches frequently occurred during missions. So, during Leonov's flight into outer space, a swollen space suit prevented the cosmonaut from returning to the ship. The American mission also experienced emergency situations. On the ship Gemini-4, when White was returning to the ship, the hatch jammed and it was not possible to close it for some time. Then there was a failure of the onboard computer and the craft could not make a normal landing.

Nevertheless, the results of the expeditions of the first people in space made it possible to make a significant breakthrough in the exploration of outer space. Many elements of flights were designed, implemented and tested. Later they formed the basis for more serious space projects. A design engineer, S. Korolev, planned to create an orbital station and organize manned flights to Venus and Mars.

The descent module of the Gemini-4 spacecraft is loaded onto the deck of the aircraft carrier after the splashdown.

The Earth's orbit

For a while, the space programs of the US and the USSR were going in different directions. The United States was engaged in the preparation of a manned flight to the Moon, and the Soviet Union was focused on the creation of an orbiting space station.

Neil Armstrong works at the module

Astronaut Buzz Aldrin on the moon

On July 20, 1969, the United States sent the spaceship Apollo 11 to the Moon. For the first time in history, the inhabitants of the Earth made a landing on another cosmic body, the Moon. They were the Americans Neil Armstrong and Buzz Aldrin. In total, American astronauts landed six times on Earth's satellite by 1972.

Saturn V carrying Apollo 11

This view of Earth was photographed from the Apollo 11 spacecraft.

Space exploration

Later, the USSR also started to develop its lunar program and carried out a number of successful lunar projects. Because of a number of factors, such as a significant lag behind the United States and inefficient organization of work, the program was only partially implemented.

Over the years it became apparent that the space race not only depletes the financial resources of rival countries but also negatively affects the scientific exploration of outer space.

In the early 70s, it became clear that the leading space nations should cooperate to achieve a more effective development of Cosmonautics.

On May 24, 1972, the USSR and the United States signed the Agreement on Cooperation in the Exploration and Use of Outer Space for Peaceful Purposes.

The Soyuz-Apollo project was set off to carry out that program. It was a joint flight of the Soviet Soyuz-19 and the American Apollo. On July 15, 1975 at 3.20 pm, Soyuz-19 was launched from Baikonur.

The Apollo spacecraft, as seen by the Soyuz crew.

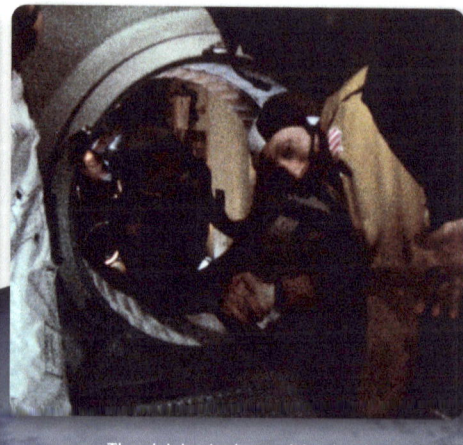

The historic handshake between Leonov and Stafford

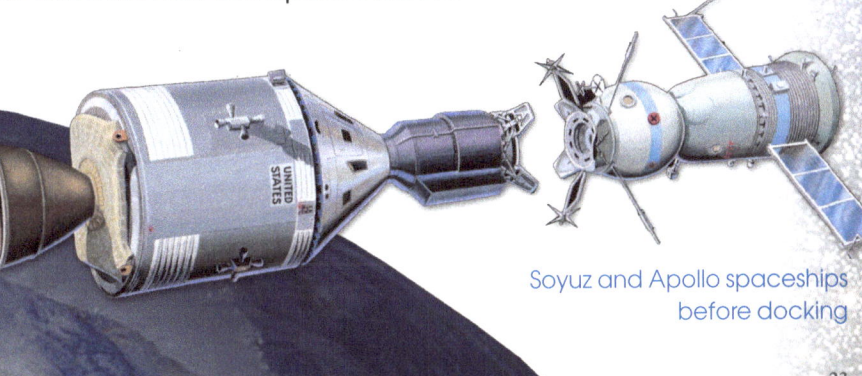

Crews of the Soviet-American mission Soyuz-Apollo spacecrafts (from left to right): Slayton, Stafford, Brand, Leonov, Kubasov.

On the same day at 10.50 pm, Apollo was launched from **Cape Canaveral**. On July 17 at 7.20 pm, the two ships were docked. After the docking, the astronauts of the American and Soviet crews started a series of technological, medical and astronomical experiments.

Experience gained during that mission was successfully used in a number of space projects, in particular, in the construction of the International Space Station.

Soyuz and Apollo spaceships before docking

LAUNCH ROCKETS

Space technology has been on a path of continuous improvement. New, more powerful and more modern launch rockets and spaceships have been created for more than half a century of space development.

At the end of the 19th century, K. Tsiolkovsky formulated the concept of human exploration of the cosmos with the help of rockets. Nowadays, a rocket remains the main vehicle for overcoming Earth's gravity and entering the cosmos.

Many types of launch rockets have been created during the decades of rocket technology development. Multi-stage rockets are the most common ones. The principle of their action is that, in the course of the flight, stages which run out of

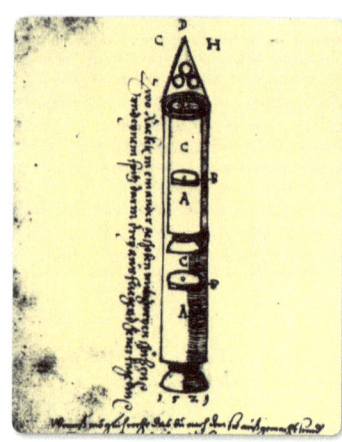

The image from 1529 shows the design of a two-stage rocket. At that time rockets were used only for fireworks.

Space Shuttle Atlantis

The Earth's orbit

Lift-off of the Dnipro launch vehicle

fuel separate from the rocket. After a stage separates from a rocket, the weight of that rocket decreases significantly. At the same time, the next stage starts to operate, and it also separates from the rocket after running out of fuel. Unlike single-stage rockets, that design enables a large amount of cargo to be sent into orbit.

Partially reusable systems have been used for a long time. Those systems included the American Space Shuttle and the Soviet Energy-Buran. The main purpose of creating those systems was to reduce the cost of manned space flights. But those systems did not justify themselves. The Energy-Buran project was canned for financial reasons after the first flight in 1988.

The Space Shuttle project was terminated in 2011. During its operation, it was not possible to improve flight safety or reduce launch costs. Also, it wasn't possible to reduce the time for preparing a launch of the next mission after returning the previous one to Earth.

Buran

Energiya-Buran system on the launch pad

The launch of a Chinese rocket from the Xichang cosmodrome

Projects similar to Space Shuttle and Buran-Energia were also developed by the European Space Agency (Hermes project) and the Japan Aerospace Exploration Agency (Hope project). But those programs were also terminated due to the inability to achieve the required performance levels. Nowadays, the most promising and partially reusable construction is the Falcon rocket.

The development of new space vehicles is ongoing.

The Vostok rocket

The Discovery shuttle on the launch pad

Space exploration

SEA LAUNCH

The effectiveness of a space rocket is very important; it depends not only on the design but also on the choice of launch site. According to calculations of navigation, the most favorable starting point is any point along the equator, where it is possible to use the Earth's rotation speed as efficiently as possible. Such a site was chosen. It was in the Pacific Ocean with coordinates 0°00' latitude 154°00' longitude, near Christmas Island. That location had been the quietest for 150 years of weather observations. There, the Sea Launch program was set up.

Sea Launch is a cosmodrome on the water. Rocket launches are made from the floating platform Odyssey. The Sea Launch program was organized by the Sea Launch Company consortium, which included Boeing, Russian Energia Corporation, Norwegian shipbuilding company Kvaerner, Ukrainian design bureau Yuzhnoye and Yuzhmash Production Association. Three-stage Zenit-3SL rockets developed by the Ukrainian consortium members were used to launch space vehicles and satellites. From 1999 to 2014, 33 successful launches of rockets were set off from that floating cosmodrome. The same number of satellites from companies of different countries and international organizations were put into orbit.

Sea Launch launch platform
Ocean Odyssey

FALCON

The Falcon family is a series of launch rockets created by the American company, Elon Musk's SpaceX. They were created on the basis of the most effective and optimal principle of two-stage liquid fuel rockets.

Nowadays Falcon rockets are widely used for launching orders by various space agencies and, above all, NASA for transporting cargo to the ISS. On September 28, 2008, the Falcon 1 launch rocket successfully delivered a payload mock-up into the required orbit.

One of the most pressing problems of space flights is their high cost. In order to reduce the cost of spacecraft launches, SpaceX undertook the development of launch rockets with a reusable first stage. Modern Falcon 9FT launch rockets use 9 Merlin engines in the first stage. Their task is to ensure the return of the first stage after the payload is put into orbit. Falcon 9FT made its first launch on December 22 2015 and put 11 satellites into orbit. After that, its first stage returned and landed successfully at Cape Canaveral, Florida. It was the first time in history when an orbital rocket stage made a soft landing after its separation.

On April 8, 2016, as a part of the SpaceX CRS-8 mission, the first stage of the Falcon 9FT rocket successfully landed on the offshore platform for the first time in the history of rocketry. On March 30, 2017, the same stage was re-launched after maintenance. It also landed successfully on the offshore platform.

Space exploration

Return of the two stages of Falcon Heavy to Earth

On February 6, 2018, Space X successfully launched the heavy Falcon Heavy rocket. It's the most powerful rocket since Saturn V. Part of the Apollo program, it was designed to lift satellites or launch spacecraft into orbit. The rocket will be able to take 63.8 tons of cargo into low Earth orbit. On the sides of Falcon Heavy, there are two accelerators, which, in fact, are the first stages from the Falcon 9FT. Therefore, both those accelerators were returned to Earth after launch.

Elon Musk's Tesla Roadster, with Earth in the background. Camera mounted on external boom.

Space exploration

MANNED FLIGHTS

To reduce the cost of manned space flights on behalf of NASA in 1971, the development of reusable space shuttles on the Space Shuttle project was declared. Five years later a similar program was initiated by the Soviet Union called Energia-Buran. Unfortunately, these systems have not fully justified the purpose of their creation yet.

Space projects similar to Space Shuttle and Buran-Energia were also developed by European Space Agency (Hermes project), Japanese Space Agency (Hope project), and Russian RSC Energia (Clipper project). But those programs were also terminated due to the inability to achieve the required performance levels.

As of 2018, there is only one program of manned flights – the Russian program based on Soyuz ships, delivered to near-Earth orbit by means of single-use rocket carriers.

Nevertheless, there are several private and state programs that have great prospects for manned space flight.

Elon Musk's American company SpaceX has won several competitions for financing the preparation of passenger flights on ships Dragon V2 (or Crew Dragon). The cargo version of Dragon has already been successfully delivered to the ISS by the Falcon 9FT launch rocket. The passenger version of the Dragon V2 will be equipped with life support and emergency rescue systems, a control panel with touch screens, as well as buttons that duplicate the main functions of the ship. Seven passengers of the Dragon V2 will be able to observe the flight from the four enlarged observation windows. To

Orion

Project of space shuttle Hermes

deliver the Dragon V2 ship into orbit, the modified first stage of the Falcon 9FT launch rocket will be used, which, like the ship itself, will be reusable. SpaceX Draco and SuperDraco engines, which were specially designed, will deal with a start, maneuver and soft landing. The first manned flight of Dragon V2 is scheduled for May 2018.

American corporation Boeing offered its version of the manned spacecraft and received financial support from NASA under the Commercial Crew Program. CST100 (or Starliner) by Boeing will consist of two modules – an instrumentation and storage compartment and a descent vehicle. The first module contains flight control systems, and will separate before entering the atmosphere while landing. The second module is equipped with a control panel and is designed for a comfortable flight of seven astronauts, as well as for cargo storage. It is planned to use the landing module up to 10 times. The first manned flight CST100 is scheduled for August 2018.

Soyuz

Space exploration

Shenzhou 5

Another manned spacecraft called Orion is being developed in the USA. It has been under development by NASA since the middle of 2000. Nowadays, Lockheed Martin continues the work on it. The purpose of creating that ship was not only the delivery of people and cargo to the ISS but also the return of American missions to the Moon and the landing of people on Mars. For the flight to the ISS, the Orion crew can include up to 6 astronauts. It was planned to send 4 astronauts in the expedition to the moon. Orion is to deliver people to the Moon for a long stay. It is the most capacious of all existing and developed manned spacecraft. When developing Orion, the engineering achievements of Apollo mission ships were used. The first manned flight of Orion is scheduled for 2021-2023.

The space agencies of Russia, India and China are also developing their manned flight programs. The first manned flight of the Chinese astronauts on Shenzhou-5 was on October 15 2003, and since then, six missions have been launched to bring people to the Chinese Tiangong-1 and Tiangong-2. The Indian Space Research Organization (ISRO) announced the launch of the national manned space program in October 2006. The spacecraft, landing module and GSLV Mark III launch rocket are still being tested. And the first manned flight of the Indian ship is scheduled for 2018-2019.

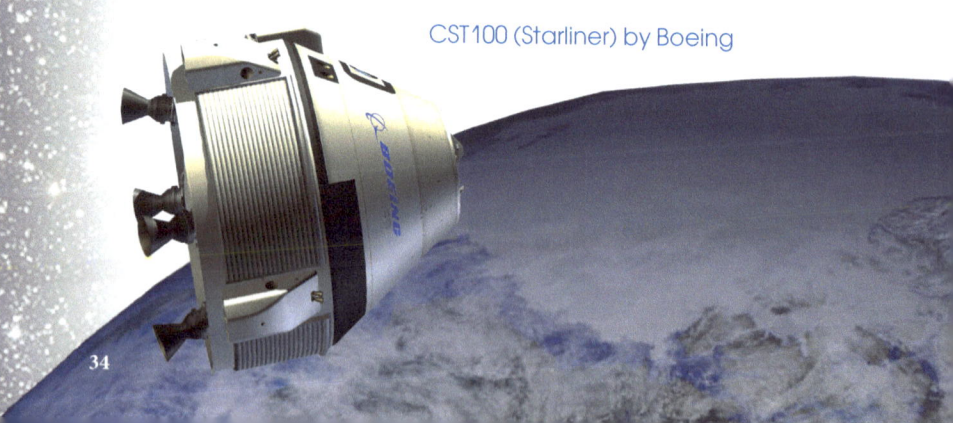

CST100 (Starliner) by Boeing

The Dragon V2 during its unveiling

SPACE TOURISM

In 1967, six years after the first successful flight of a human being into space the ideas of space tourism and the commercialization of space flights were discussed for the first time. But at that time those ideas weren't appreciated. However, at the end of the 20th century, space tourism began to be actively discussed in scientific and business groups. The first space tourist was supposed to be a teacher Christa McAuliffe, who died at the launch of the shuttle Challenger in 1986. After that accident, the US government passed a law banning non-professionals from flying into space.

SpaceShipTwo «Unity»

But there was no ban on flights in Russian ships. In 1990 and 1991, cosmonauts Toyohiro Akiyama (Japan) and Helen Sharman (Great Britain) flew to the Mir space station. The first space tourist was American multimillionaire Denis Tito. He paid his own 20 million dollars for the flight in 2001. After Denis Tito, tourist flights to space became popular among rich people around the world. Today there are seven space tourists who have visited the ISS, one of them is the American billionaire Charles Simonyi. He has already been there twice.

Today, tourists get to the ISS as a part of missions on Soyuz ships, but in the future, with the commissioning of Dragon V2, CST100, Orion ships, the flights of tourists to near-Earth orbit will not only become more frequent but also less expensive.

Some commercial tourist programs do not provide for flights to the ISS. For example, the British company Virgin Galactic offers to get to a height of 100 km on SpaceShipTwo ships, stay in zero gravity for about 4 minutes, and then land back on Earth. Suborbital programs are also being developed by Blue Origin and Xcor Aerospace. Roskosmos promises tourist overflights around the Moon on Soyuz ships in the very near future. Elon Musk intends to offer the same service on Dragon V2 ships. American company Bigelow Aerospace plans to create something like a hotel for space tourists. It will be a complex of inflatable modules, two of which were successfully delivered into orbit by Ukrainian Dnepr launch rockets in 2006 and 2007. The third module was docked to the ISS in 2016 and is now being tested.

Xcor Rocket Racer landing after its first flight

© Alan Radecki Akradecki

Space exploration

CARGO SPACECRAFT

In addition to satellites and manned spacecraft, unmanned cargo ships are regularly sent into the Earth's orbit. Their task is the delivery of fuel, equipment, provisions, devices for scientific research and consumables to the orbital station. The first cargo ship was the Soviet transport ship TKS; it was developed together with the first Salyut-1 orbital station in 1971. It was able not only to deliver goods into the orbit but also to return them to Earth. In particular, its landing capsule returned photos from outer space. Today, ships from several countries, including Russia, the United States, Japan and the European Union carry cargo to the ISS. In April 2017, the first Chinese Tianzhou transport vehicle flew into orbit; it docked with the Chinese space laboratory Tiangong-2.

The Russian automatic unmanned cargo spacecraft Progress was created on the basis of the Soyuz spacecraft. From the very beginning, it was decided to make it one-time and the first ships of that series returned to Earth. They burned up in the atmosphere. Then, with the advent of additional docking sets at the Salyut stations, the empty Progress "trucks" began to be used as additional station rooms.

In March 2008, the first Automated Transfer Vehicle was sent into orbit by the European Space Agency. Its task was to deliver cargo to the ISS, adjust its orbit with the help of its engines and free the station from waste materials and debris. After staying in orbit, European "trucks" were sent to Earth and burned up in the atmosphere. In total, five ATV ships were launched, each

SpaceX Dragon commercial cargo

Cygnus

of which was given a unique name in honor of an outstanding writer or scientist — "Jules Verne", "Johannes Kepler", "Edoardo Amaldi", "Albert Einstein" and "Georges Lemaitre".

HTV-1

The Japanese aerospace research agency JAXA launched its unmanned cargo vehicle HTV in 2009. After unloading, that ship undocks and then burns up in the atmosphere, just as do Russian and European ships.

NASA entrusted the development and creation of unmanned rockets in the USA to private companies. During a contest, two of them were chosen — SpaceX with the Dragon ship and Orbital Sciences Corporation with the Cygnus ship.

The unmanned cargo ship Cygnus has similar design and functions to the European ATV.

The cargo ship Dragon is significantly different from all other ships delivering goods into orbit. It's the only reusable unmanned ship nowadays, which delivers "parcels" with the ISS to Earth. In the ship, there are results of biotechnological research, other scientific experiments and materials for cognitive programs. In addition, unlike other returned ships, i.e. Apollo, Soyuz, ships developed by CST-100, Orion and Federation, the Dragon returns to Earth all the components needed to start: engines, tanks, batteries and other equipment. That function makes it unique.

ATV
Progress

ORBITAL STATIONS

After man's flight into space, the development of near-Earth space and the construction of orbital bases began. So far, they are relatively small bases, used mainly for scientific purposes.

To fulfill complex research tasks in space at the present stage of technology development, individual satellites or short-term space missions are no longer sufficient. To ensure the stay of an astronaut or a group of astronauts, a complex orbital station is needed, equipped with all the necessary scientific devices, it also must have sufficient living conditions.

In fact, Orbital stations are also artificial satellites of the Earth. But in their structure, functionality and significance, they are much superior to satellites in the traditional sense. Orbital stations, as a rule, include several modules: for scientific research, for life support of the crew and a transport ship.

Proparullon for the launch of the first orbital station Salyut-1

Launch of the Saturn V rocket with the US space station Skylab

The possibility of creating an orbital station resulted from many years of research. The most important element in the orbital station is the possibility of docking it with a transport ship. To develop that process in the late 1960s, the United States and the USSR initiated several missions.

The first orbital station was launched by the Soviet Union on April 19 1971. The station was called Salyut. On June 6, the crew was delivered to it within the framework of the first long-term space expedition. Three cosmonauts conducted various studies and tests of the onboard systems of the station in various modes for 23 days. They were observing the Earth's surface and meteorological phenomena in the Earth's atmosphere. In May 1973, the US launched its station called Skylab. Expeditions at that station have established several records for the duration of the astronauts' stay in space. The tasks of the expeditions at Skylab

Skylab

Space exploration

Astronaut Owen Garriott works next to Skylab's manned solar space observatory, 1973

included the study of the Earth's natural resources, observations of the Sun and technological experiments.

The Soviet station Mir started the modern concept of development of orbital stations. It was put into orbit in early 1986. In 10 years, the expedition turned into an advanced research and development complex.

The station Mir embodied the modular principle of construction. It involves the creation of a fully functional basic module, to which other modules can later be mounted and dismounted, depending on the current tasks. The same principle, as well as many technologies worked out during the construction of the Mir station, are now used for the de-

Interior of the Salyut-4 orbital station

velopment of the International Space Station project.

During the operation of the MIR station, 104 cosmonauts from 12 countries visited it. More than 23,000 experiments were conducted at the station.

In the late 90s, due to frequent disruptions in the station's systems, it was decided to stop its operation. On March 23, 2001, Mir station was sunk in the southern part of the Pacific Ocean.

Docking of the Atlantis shuttle and the Mir station, 1995

The Mir station in 1997

INTERNATIONAL SPACE STATION

In 1984, the United States initiated a project to build a manned space station. In the process of the work, Canada, Japan and the European Space Agency joined the project. By the beginning of the 1990s, it had become clear that without closer international cooperation, the project could not be implemented because of its complexity and high cost. The USSR was considered to be one of the important potential partners for the project implementation. The Soviet Union had a rather rich experience in the creation of a modular space station. In the USSR the Mir-2 orbital station project was terminated due to lack of funds. Therefore, the cooperation of the US and the USSR in this project was beneficial for each side.

ISS insignia

On June 17 1992, an agreement on joint exploration of outer space was concluded between the US and the Russian Federation (the successor of the USSR). Several programs were developed, during which the International Space Station (ISS) project was set off. On 1 November 1993, the RSA and NASA signed a work plan for the International Space Station. In 1996, the station's configuration was approved. Based on the project, the station was divided into two main

The Zarya module - The first block of the ISS

The Earth's orbit

Unity module

segments: Russian and international (USA, Canada, Japan, Italy, Brazil, and European Space Agency).

On November 20, 1998, Russia set into the orbit the first block of the ISS, the functional-cargo block Zarya. After 17 days, the shuttle Endeavor docked the ISS Unity module. This was the practical beginning of one of the largest space projects in history.

On November 2, 2000, the crew of the first expedition arrived at the ISS on the Soyuz TM-31 transport ship.

STATION POWER SUPPLY

The only source of electricity on the ISS is sunlight. The station is equipped with several solar panels. Rechargeable batteries feed the station while it is in the shadow of the Earth and solar panels are not functioning. When the station turns back to the Sun, those batteries are charged.

Since 2003, after the Columbia shuttle disaster, the United States has suspended shuttle launches. Supplying the station with materials for research and vital activity maintenance was carried out only by Russian ships. ISS long-term expe-

Crew of the first expedition to the ISS (from left to right): S. Krikalyov, U. Sheppard, Yu. Gidzenko

Space exploration

ISS in 2000

dition crew was reduced from three to two. In 2006, after the resumption of shuttle flights, the German cosmonaut Thomas Reiter arrived at the ISS. The crew of the station consisted of three people again.

In 2008, the European Space Agency launched the ATV transport ship, and in 2009 the Japanese cargo ship flew to the station for the first time. The appearance of those craft allowed increasing the volume of cargo delivered to the ISS, and as a result, to increase the crew of a long-term mission to six people.

In its 20 years of existence, the station has grown to 23 modules.

In total, 14 countries take part in the ISS project: Belgium, Germany, Denmark, Spain, Italy, Netherlands, Norway, Sweden, France, Switzerland, within the framework of the programs of the European Space Agency; Canada, Russia, the United States, Japan.

ISS in 2011

SCIENTIFIC RESEARCH IN SPACE

The scientific part of the ISS program includes conducting experiments at the station under vacuum and cosmic radiation conditions. The main research is carried out in the fields of biology, physics, astronomy and meteorology. The station has three scientific modules: the American Destiny, the European Columbus, and the Japanese Kibo.

Destiny module

Joint research on the ISS has been greatly improved. For example, Russia, together with ESA, is carrying out a series of medical and biological experiments.

The American scientific segment on the ISS is mainly concerned with biological research. The Japanese mission deals with environmental issues: global warming, desertification of the Earth's surface, reduction of the ozone layer. The research program of the European Space Agency is the most diverse. On the Columbus scientific module is a laboratory for conducting biological experiments, for studying liquid physics, a solar observatory, equipment for performing experiments on crystallization of proteins, and a module for experiments in physiology.

Columbus module

SATELLITE SYSTEMS

Nowadays, satellites in the Earth's orbit are an integral part of life support. A huge number of industries depend on the work of orbiting satellites.

An important direction in the development of outer space was the development and launching of several types of artificial satellites into orbit. Their tasks include the provision of a number of processes occurring on Earth. Such satellites primarily include communication satellites, meteorological satellites, which help to observe weather conditions on the planet; reconnaissance satellites that perform a wide range of functions – from mineral exploration and environmental monitoring to the fulfillment of military tracking tasks.

Launch of the Chinese meteorological satellite Fengyun-2, December 23, 2008

The Soviet meteorological satellite Kosmos-122

The Earth's orbit

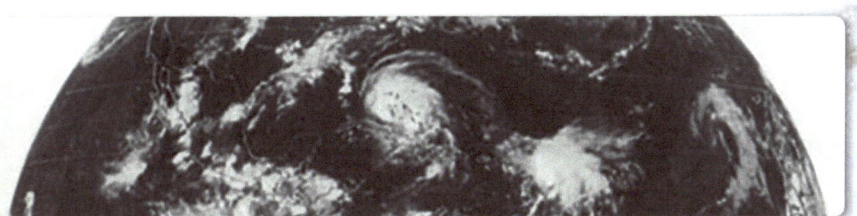

This picture was taken by the SMS-2 weather satellite in September 1979. It clearly shows two powerful hurricanes moving to the eastern coast of North America. These hurricanes were given the names David and Frederick.

Thanks to the timely location finding of the impending elements, it was possible to warn the population and take all necessary measures. When the hurricanes hit the Caribbean and the American coast, their destructive effect was minimized.

METEOSATELLITES

The creation of the first meteorological satellites led to the emergence of a separate field of science – satellite meteorology. Meteosatellites watch over the Earth's surface, its temperature, cloud and snow cover. Data from meteorological satellites allows tracking weather anywhere in the world. It's especially important for predicting such phenomena as a hurricane, a snowstorm, a storm.

The first meteorological satellite was launched by the USA in April 1960. It was called TIROS-1 and its task was exclusively to monitor the weather.

In the USSR, the first meteorological satellite Kosmos-122 was launched into the orbit in June 1966.

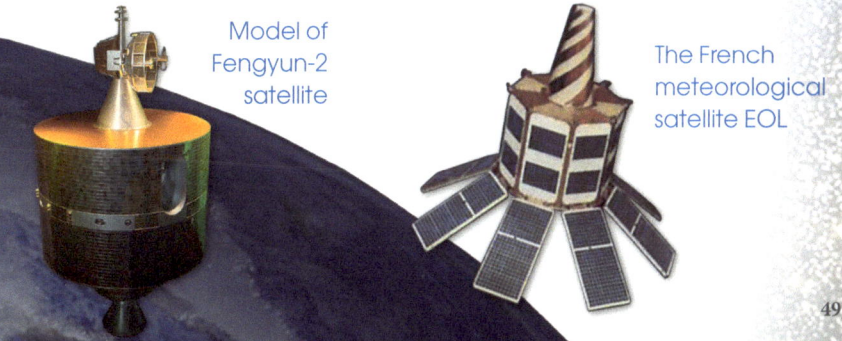

Model of Fengyun-2 satellite

The French meteorological satellite EOL

Space exploration

SATELLITE CONNECTION

One of the most important types of radio communication is satellite communication. A satellite which is set in the orbit performs linkage between Earth stations that can be spaced a great distance apart.

Echo-1

The idea of an orbiting radio repeater satellite was expressed back in 1945 by the English scientist Sir Arthur Clarke. But the first practical studies in the field of civil satellite communication were conducted only in the late 50s.

On August 12, 1960, the United States put into orbit an Echo-1 inflatable balloon covered with a metallized shell. It became the first communication satellite in orbit. Later, that type of communication satellite wasn't further developed due to its low efficiency. All communication satellites are active responders, which have equipment for receiving, transmitting and amplifying the signal.

In August 1964, 11 countries established the international satellite communications organization Intelsat. The purpose of that organization was to create a satellite network to meet the growing need for high-quality and low-cost communications.

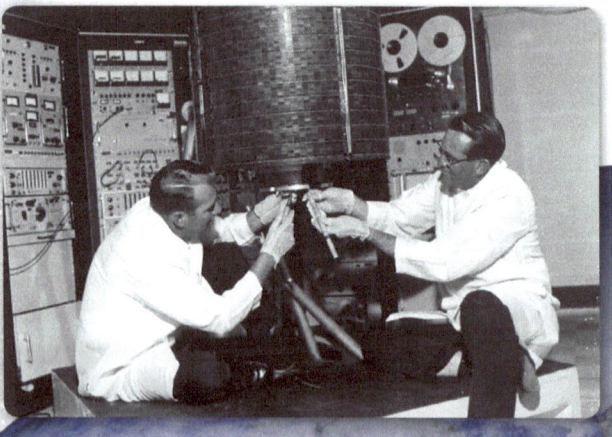

INTELSAT I
Early Bird

The Earth's orbit

Communications satellite Syncom-1

Milstar Satellite Communications System

The first result of Intelsat's work was the launch of ComSat's commercial communications satellite EarlyBird into the orbit. That satellite could provide up to 240 communication channels, which at that time was a kind of technological breakthrough. The satellite provided communication between the station in the US and one of the stations in Europe – in the UK, France or Germany.

In 2015, Space X announced the launch of the project to create a system of satellites to provide a cheap and reliable communication channel. According to the project authors, 12,000 satellites in near-earth orbit should provide high-quality communication with any point of the Earth. In February 2018, Space X launched Tintin A and Tintin B satellites to the orbit, which became the first satellites of this system.

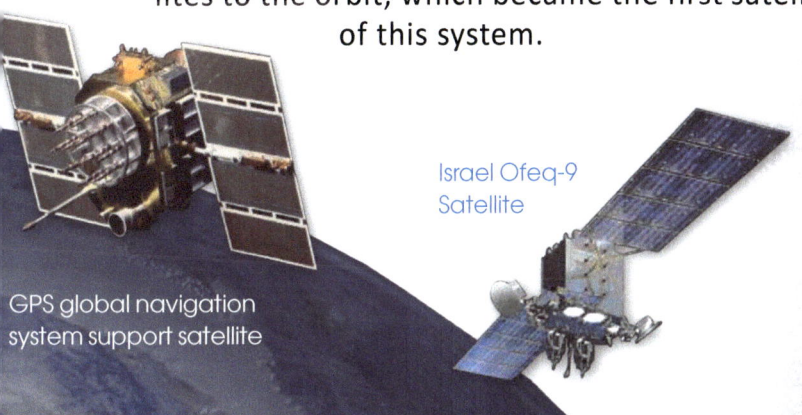

GPS global navigation system support satellite

Israel Ofeq-9 Satellite

SCIENTIFIC SATELLITES

Satellites are also used for scientific research. They are used both for studying the Earth and the processes taking place on it and for studying the Cosmos. The greatest interest for scientists is represented by astronomical satellites – stations that explore the Cosmos from orbit. The terrestrial atmosphere dissipates the radiation of cosmic objects. Therefore, in order to obtain the most accurate data, it is advisable to observe the Cosmos from orbit. Space telescopes, as a rule, are equipped with devices for obtaining gamma, X-ray and ultraviolet radiation, and devices for transmitting information.

Lyman Spitzer

For the first time, the idea of an orbiting telescope was described in 1923 by Hermann Oberth. And in 1946, an American scientist Lyman Spitzer published an article in which he argued for studying the cosmos in such a manner. Since that time, the development of space astronomy has been developing. In 1962, the United Kingdom launched an orbiting telescope to observe the Sun. NASA included the creation of an orbiting telescope

Orbiting Astronomical Observatory «OAO-1»

The Earth's orbit

Orbiting Astronomical Observatory «OAO-2»

Launch of the Atlas carrier rocket with OAO-1

into its space program, and in 1966 it launched the Observatory OAO-1 into the cosmos. The launch was successful, but the project failed. Three days after entering orbit, the batteries of the observatories stopped working. The launch of the second observatory OAO-2 was much more successful. Launched into orbit in 1968, the station operated until 1972. With the help of OAO-2, observations of ultraviolet radiation from stars and galaxies were carried out.

Space exploration

HUBBLE SPACE TELESCOPE (HST)

The results of the OAO-2 and project showed how important and useful it is to make astronomical observations from orbit. In 1968, NASA developed a program to build an orbiting telescope with a mirror diameter of 3 meters. According to the program, the telescope not only had to be put into the orbit but, it was necessary to launch manned flights to it to maintain it.

The initial stage of the space telescope assembly

The development of the orbiting telescope project did not go quite smoothly. Its cost is many times higher than the budget for the construction of any terrestrial telescope. In order to reduce costs, it was decided to reduce the diameter of the telescope's mirror to 2.4 meters. The European Space Agency has joined the project, having undertaken the financing of some of the works. Full-fledged work on the project began only in 1978, with the planned launch date for 1983. The telescope was named after the outstanding astronomer Edwin Hubble. The high tech-

Grinding of Hubble's primary mirror at Perkin-Elmer, March 1979

Launch of the Discovery shuttle with the Hubble telescope on board

nological specification of the project did not allow investing by the deadline and the launch date was postponed to 1986. The disaster of the Challenger, which was supposed to bring the satellite into orbit, suspended the program of using space shuttles. Therefore, the launch of the telescope was postponed for some time.

The observatory had to be left on Earth and special storage conditions had to be created for it. Some of the on-board sys-

The Hubble Space Telescope as seen from the departing Space Shuttle Atlantis

Space exploration

Astronauts Musgrave and Hoffman install corrective optics

tems were put into operation. In reality, that delay made it possible to introduce a number of improvements and upgrades into the design of the telescope. In particular, solar cells were replaced with high-efficiency batteries; the telescope control software was also added. In 1988, NASA resumed the program of using space shuttles and the launch of the telescope into orbit became possible.

On April 25, 1990, the Discovery shuttle brought the Hubble telescope to its desired orbit.

On December 2, 1993, the Endeavour shuttle delivered a crew to the Hubble Space Telescope for maintenance and troubleshooting. That mission became one of the most complicated

Hubble in the cargo compartment of the shuttle

in the history of space exploration. The cosmonauts' task was not only to establish corrective optics but also to replace solar batteries, gyros of the guidance system and update the computer system. The shuttle spent ten days in the orbit, during which five spacewalks were carried out by cosmonauts.

During the 1999 expedition, the telescope was significantly upgraded. The onboard computer was replaced by a more powerful computer which was resistant to radiation. Two new modules were also put into operation enabling it to explore deep space. After the last expedition in 2006, according to calculations of specialists, the telescope had to work without maintenance until 2014. After that, it was planned to replace Hubble with a new generation orbital telescope. But in 2016, the Hubble program was officially extended for five years, until 2021.

While working in orbit, the Hubble telescope has made more than 1 million images of various celestial objects. Its technical capabilities were used by about 4000 astronomers from all corners of the Earth.

The most famous photos made by the Hubble telescope

Space exploration

INTELLIGENCE SATELLITES

Despite the exclusively peaceful nature of research and exploration of the Cosmos, the arms race between the USSR and the United States was reflected in space technologies at some point.

On February 28, 1959, the United States launched the first satellite-photo-scanner, which was supposed to conduct photographic observations outside the territory of China and the USSR. In May of the same year, work on the reconnaissance satellite also began in the USSR. As a result, the Zenit-2 photo reconnaissance complex was created and was adopted in 1964. Soviet and American reconnaissance satellites were quite primitive devices. The shooting was carried out on film; after that the capsules with the film were sent to Earth. Subsequently, with the development of digital technologies, information from satellites began to be transmitted in the form of an encrypted digital signal, which simplified and accelerated its processing.

The Vela-5A/B Satellite in its cleanroom. The two satellites, A and B, were separated after launch.

Like all military technologies, programs of reconnaissance satellites are shrouded in secrecy. But the number of developments of such devices can be figured out due to the fact that at present in the Earth's orbit there are about 2000 spent reconnaissance satellites that were produced in the USSR.

U.S. Lacrosse radar spy satellite under construction

SPACE AGENCIES

The development of space is one of the most urgent and promising tasks of mankind. Therefore, it is not surprising that since the launch of the first artificial Earth satellite this subject has received so much attention all over the world. As a result, national space agencies began to appear one after another in the mid-1980s. Today, the most significant work in the exploration of outer space is being conducted by the United States. However, large organizations such as the European Space Agency (ESA), the Japan Aerospace Exploration Agency, the China National Space Administration (CNSA), the Indian Space Research Organization (ISRO), invest more and more time to deepen the study of the cosmos. In many countries – Canada, Spain, the Netherlands, Germany, South Korea, Israel and others – numerous research institutes work on the problems of space exploration. Even developing countries such as Peru, Bangladesh, Nigeria, Thailand, Azerbaijan and Ecuador have organized their space agencies to promote research, working with larger agencies around the world.

www.ingramcontent.com/pod-product-compliance
Lightning Source LLC
Chambersburg PA
CBHW040237220526
45473CB00001B/278